8th International Conference on Stem Cell Engineering 2022

AA002609

Cambridge, Massachusetts, USA
7 – 9 October 2022

ISBN: 978-1-7138-6320-5

8th International Conference on Stem Cell Engineering 2022

Cambridge, Massachusetts, USA
7 – 9 October 2022

AIChE
The Global Home of Chemical Engineers

Printed from e-media with permission by:

Curran Associates, Inc.
57 Morehouse Lane
Red Hook, NY 12571

Some format issues inherent in the e-media version may also appear in this print version.

Printed with permission by Curran Associates, Inc. (2023)

For permission requests, please contact AIChE
at the address below.

AIChE
120 Wall Street, FL 23
New York, NY 10005-4020

Phone: (800) 242-4363
Fax: (203) 775-5177

www.aiche.org

Additional copies of this publication are available from:

Curran Associates, Inc.
57 Morehouse Lane
Red Hook, NY 12571 USA
Phone: 845-758-0400
Fax: 845-758-2633
Email: curran@proceedings.com
Web: www.proceedings.com

TABLE OF CONTENTS

ORAL

POSTER

Integrated Generation of Induced Pluripotent Stem Cells in a Low-Cost Device

- Conference: International Conference on Stem Cell Engineering
- Year: 2022
- Proceeding: 8th International Conference on Stem Cell Engineering
- Group: General Submissions
- Session:

  Oral

Authors:
Lei, Y. - Presenter, University of Nebraska - Lincoln
Wang, O., University of Nebraska-Lincoln

Human induced pluripotent stem cells (iPSCs) have unlimited proliferation capability and potential to differentiate into all somatic cells. Their derivatives contain patientsâ€™ genetic information and can model many diseases. Additionally, derivatives of patient-specific iPSCs induce minimal immune rejection *in vivo*. With this unique combination of properties, iPSCs open the avenue to personalized medicine including personalized drug screening, toxicity test, cell therapy and tissue engineering. However, the further advance of iPSC-based personalized medicine is currently limited by the difficulty to generate iPSCs for large populations and at affordable cost. We here report a low-cost device to address this challenge. The device allows the entire bioprocess for generating high quality and quantity of iPSCs for one patient to be done automatically within a closed conical tube without cell passaging. Additionally, iPSCs can be further differentiated into somatic cells in the device. Thus, the device also allows integrated iPSCs generation, expansion and differentiation to produce any somatic cell types. This device can be made in large quantities at low cost for manufacturing iPSCs (and their derivatives in necessary) for large populations at affordable cost. It will significantly advance the iPSCs-based personalized medicine.

1) Integrated generation of induced pluripotent stem cells in a low-cost device. Biomaterials 189, 23-36

2) Manufacturing human pluripotent stem cell derived endothelial cells in scalable and cell-friendly microenvironments. Biomaterials science 7 (1), 373-388

3) Engineered microenvironment for manufacturing human pluripotent stem cell-derived vascular smooth muscle cells. Stem cell reports 12 (1), 84-97

Cybersecurity Concerns in Pharmaceutical Industry

- Conference: International Conference on Stem Cell Engineering
- Year: 2022
- Proceeding: 8th International Conference on Stem Cell Engineering
- Group: General Submissions
- Session:

  Oral

Glioblastoma (GBM), the most common type of primary brain tumor, is characterized by high mortality rate, short lifespan, and poor prognosis with a high tendency of recurrence. Functional therapeutics, including PRMT5 inhibitors, radiosensitizers, and emerging chimeric antigen receptor (CAR)-T immunotherapy, have been developed to treat GBM. However, the existence of physiological blood-brain barrier (BBB) or blood-brain-tumor barrier has impeded the efficient delivery of such promising therapeutics into the brain and limited their therapeutic efficacy. Given the native ability of neutrophils to cross BBB and penetrate the brain parenchyma, here we tested the therapeutic concept that neutrophils could be engineered with synthetic CARs to specifically target GBM and effectively deliver chemo-drugs to brain tumor as a novel dual chemoimmunotherapy for the first time. Primary neutrophils are short-lived and resistant to genetic modification. Therefore, we genetically engineered human pluripotent stem cells with different chlorotoxin (CLTX) CARs and differentiated them into functional CAR-neutrophils. As compared to CAR-natural killer (NK) cells, systemically administered hPSC-derived CLTX CAR-neutrophils significantly reduced tumor burden in xenograft mouse models and extended their lifespan, suggesting superior abilities of neutrophils in crossing BBB and penetrating GBM xenograft in mice. We also loaded hypoxia-activated prodrug tirapazamine (TPZ) into CAR-neutrophils using silica nanoparticles with rough surfaces (R-SiO_2-TPZ) and demonstrated their enhanced antitumor activities in xenograft mouse models, serving as a novel dual chemoimmunotherapy against GBM. Our results established that CAR neutrophil-mediated drug delivery may provide an effective and universal strategy for specific targeting of solid tumors.

In Vitro Culture Expansion Shifts the Immune Phenotype of Human Adipose-Derived Mesenchymal Stem Cells

- Conference: International Conference on Stem Cell Engineering
- Year: 2022
- Proceeding: 8th International Conference on Stem Cell Engineering
- Group: General Submissions
- Session:

  Oral

Authors:
Li, Y. - Presenter, Florida State University
Jeske, R., Florida State University
Yuan, X., FAMU-FSU College of Engineering
Bunnell, B., Tulane University

Human mesenchymal stem cells (hMSCs) are widely used in regenerative medicine due to their differentiation abilities, secretion of trophic factors, and regulation of immune responses in damaged tissues. Due to the limited quantity of hMSCs isolated from bone marrow (BM), adipose tissue-derived hMSCs (hASCs) are considered a promising alternative. However, differences have been observed for hASCs in metabolic characteristics and response to *in vitro* culture stress compared to BM-hMSCs. In particular, the immune phenotype and immunomodulation of hASCs, remains unknown. This study thoroughly assessed the changes in metabolism, redox cycles, and immune phenotype of hASCs during *in vitro* expansion. In contrast to BM-hMSCs, hASCs did not respond to culture stress significantly during expansion as limited cellular senescence was observed. Notably, hASCs exhibited the increased secretion of pro-inflammatory cytokines and the decreased secretion of anti-inflammatory cytokines after extended culture. The NAD+/NADH redox cycle and other metabolic characteristics associated with aging were relatively stable, indicating that hASC functional decline may be regulated through an alternative mechanism rather than NAD+/Sirtuin aging pathways as observed in BM-hMSCs. Furthermore, transcriptome analysis by mRNA-sequencing revealed the upregulation of genes for pro-inflammatory cytokines and the downregulation of genes for anti-inflammatory cytokines for hASCs at high passage. Proteomics analysis indicated key pathways (e.g., tRNA charging, EIF2 signaling, and protein ubiquitination pathway) that may be associated with the immune phenotype shift of hASCs. Together, this study advances our understanding of the metabolism and senescence of hASCs and may offer vital insights for the biomanufacturing of hASCs for clinical use.

Bio-Engineered Liver Organoids for Drug Screening and Cell Replacement Therapy

- Conference: International Conference on Stem Cell Engineering
- Year: 2022
- Proceeding: 8th International Conference on Stem Cell Engineering
- Group: General Submissions
- Session:

  Oral

Authors:
Pettinato, G. - Presenter, Beth Israel Deaconess Medical Center
Fisher, R. A., Beth Israel Deaconess Medical Center
Zhenghui, G. J., BIDMC

Introduction

Engineered tissue represents a great promise for treating liver diseases. Differentiated liver organoids have shown to be an interesting alternative to the native liver-derived hepatocytes. Controlling the progression of differentiation in live cells could have important implications for obtaining fully functional terminally differentiated liver organoids. In order to monitor the progression of differentiating organoids consisting of HLA-Class II knockout hiPSCs interlaced with endothelial cells toward functional liver organoids, we have complemented the conventional real-time PCR measurements of hepatic markers and coagulation factors with the noninvasive Raman spectroscopy.

Methods

Generation of human embryoid bodies (hEBs) using HLA-Class II knockout hiPSCs interlaced with endothelial cells was achieved using ROCKi/Spin-free technology we developed earlier. Raman spectra were collected in live organoids using a Raman spectroscopy system consisting of a 785 nm excitation diode laser and a portable Raman spectrometer. We also performed clinically relevant assays for measuring hepatic function *ex vivo*.

Results

Changes in Raman spectra of organoids over the course of 16 days, spanning four stages of differentiation, exhibited peaks that can be associated with CER1, AFP, and creatinine (Fig. 1a-c). Gene expression of hepatic markers and coagulation factors were detected by real-time PCR, showing the maturation of liver organoids (Fig. 1d).

Conclusions

Real-time monitoring of differentiating hiPSC organoids represents an excellent tool to assess the differentiation progress in live cells. Our results demonstrated the potential of Raman spectroscopy to complement conventional techniques providing information on hepatic differentiation noninvasively. This technique could be important for successful clinical translation.

An Engineered Multicellular Stem Cell Niche to Study Skeletal Muscle Physiology and Function

- Conference: International Conference on Stem Cell Engineering
- Year: 2022
- Proceeding: 8th International Conference on Stem Cell Engineering
- Group: General Submissions
- Session:

  Oral

Authors:
Mashinchian, O. - Presenter, Nestlé Institute of Health Science, Nestlé Research
Fate decisions in the embryo are controlled by a plethora of microenvironmental interactions in a three-dimensional niche. To investigate whether aspects of this microenvironmental complexity can be engineered to direct myogenic human induced pluripotent stem cell (hiPSC) differentiation, we screened cell types present in the developmental or adult stem cell niche in heterotypic suspension embryoids. We identify the combination of embryonic endothelial cells and fibroblasts as highly permissive for myogenic specification of hiPSCs. Combined with Wnt and FGF pathway mediated lineage-priming, embryoid differentiation allows for the isolation of >99% pure Pax7 positive skeletal muscle progenitors within a two-week time-window. Myogenic differentiation of hiPSCs in heterotypic embryoids relies on a specialized structural microenvironment and activates the MAPK, PI3K/AKT and Notch pathways. Embryoid derived myogenic cells reactivate after repeated injury and, compared to adult human myoblasts, display enhanced fusion to muscle fibers and lead to stronger muscles when transplanted into a mouse model of Duchenne muscular dystrophy. Altogether, we provide a highly efficient and scalable suspension-based protocol for 3D derivation of Pax7 positive myogenic progenitors from hiPSCs.

Genetically Engineered Human Pluripotent Stem Cells Utilizing Multi-Step Automated Differentiation Networks for Development of Three-Dimensional Liver Organoids

- Conference: International Conference on Stem Cell Engineering
- Year: 2022
- Proceeding: 8th International Conference on Stem Cell Engineering
- Group: General Submissions
- Session:

  Oral

Authors:
Mishra, D. - Presenter, MIT
Weiss, R., Massachusetts Institute of Technology
Wang, L., Institution of Process Engineering, Chinese Academy of Sciences, Beijing
Perry, E., MIT
Baudy, A., Merck & Co., Inc.
Roesner, J., Merck & Co., Inc.
Pacchione, S., Merck & Co., Inc.
Su, M., Merck & Co., Inc.
Summers, N., MIT
Gibson, C., Merck & Co., Inc.
Three-dimensional (3D) human stem cell-derived systems allow recapitulation of architecture, composition, and physiology of organs. Organoids may be formed by combining individually differentiated cell populations or via introduction of external cues to human induced pluripotent stem cells (hiPSCs) to mimic stages of human development. However, such approaches may limit intercellular interactions or faithful mimicry of a mature organâ€™s structure. Here, we present an approach to genetically engineer hiPSCs with transcriptional and post-transcriptional regulatory networks to achieve multi-step differentiation in a cell-specific, semi-automated manner to form 3D liver organoids. The system relies on two independently operating networks, the first is small-molecule inducible heterogeneous expression of GATA-binding protein 6 (Gata6) to initiate germ layer formation and is capable of directing formation of 3D vascularized liver organoids. The second network operates post-transcriptionally, coupling miRNA sensing of miR-122a-5p and restriction endoribonucleases (ERNs) to autonomously express protein payloads in a hepatic-lineage cell-conditional manner. These engineered iPSCs were grown for thirty days to form 3D liver organoids and when compared to a single-stage network, exhibited 60% increase in urea production, 36% increase in albumin production, and an order of magnitude increase of Cyp3A4 enzyme function while maintaining vasculature and cell compositional diversity. Collectively, our two-stage genetic network approach enabled selective differentiation, maturation, and conditional payload expression for hepatocyte development. We anticipate

systems such as ours utilizing cell-specific miRNA sensors and ERN controlled payloads will allow selective differentiation of germ layers and progenitors as well as cell maturation across many existing organoid models.

Microbial Surfactants As Green Emulsifiers for Skincare Cosmetic Nano-Formulations

- Conference: International Conference on Stem Cell Engineering
- Year: 2022
- Proceeding: 8th International Conference on Stem Cell Engineering
- Group: General Submissions
- Session:

  Oral

Authors:
Kapila, S. - Presenter, Birla Institute of Technology and Science Pilani -
Nanoemulsions have gained special attention in the cosmetic industry. In recent years, there has been a paradigm shift from synthetic to natural-based products because of changing lifestyles, increasing awareness and incomes, and accessibility. One of the natural motivations for this quest is finding suitable emulsifier candidates with negligible side-effects that are sourced from plants or microbes, which can serve as viable replacements to the erstwhile used synthetic surfactants.

This research draws attention towards the preparation of new cosmetic surfactants with safer, biocompatible, biodegradable, and naturalâ€□based properties The collective knowledge churned out through a comprehensive literature survey revealed that certain microbial surfactants such as Mannosylerythritol lipids (MEL), surfactin, Rhamnolipids, sophorolipids, trehalolipids, biosurfactants produced using *Bacillus subtilis,* illustrate important properties, which consequently, can be efficiently used in cosmetic formulations such as eyeshades, lipsticks, soap, powders, etc. Rhamnolipid stands as a potential competitor when essentially compared to an existing widely used surfactant, quillaja saponin which has been used before in the delivery of Vitamin E, an important bioactive compound in cosmetic products.

But it is rather whimsical to call the natural alternatives completely safe and more sustainable. High production cost and confined structural variety are some limitations in further applications of biosurfactants. Therefore, exploring the molecules from renewable sources in order to supplant the raw materials derived from petroleum, has become the need of the hour.

Engineer CAR-Neutrophils from Human Pluripotent Stem Cells for Targeted Chemoimmunotherapy Against Glioblastoma

- Conference: International Conference on Stem Cell Engineering
- Year: 2022
- Proceeding: 8th International Conference on Stem Cell Engineering
- Group: General Submissions
- Session:

  Oral

Authors:
Bao, X. - Presenter, Purdue University
Glioblastoma (GBM), the most common type of primary brain tumor, is characterized by high mortality rate, short lifespan, and poor prognosis with a high tendency of recurrence. Functional therapeutics, including PRMT5 inhibitors, radiosensitizers, and emerging chimeric antigen receptor (CAR)-T immunotherapy, have been developed to treat GBM. However, the existence of physiological blood-brain barrier (BBB) or blood-brain-tumor barrier has impeded the efficient delivery of such promising therapeutics into the brain and limited their therapeutic efficacy. Given the native ability of neutrophils to cross BBB and penetrate the brain parenchyma, here we tested the therapeutic concept that neutrophils could be engineered with synthetic CARs to specifically target GBM and effectively deliver chemo-drugs to brain tumor as a novel dual chemoimmunotherapy for the first time. Primary neutrophils are short-lived and resistant to genetic modification. Therefore, we genetically engineered human pluripotent stem cells with different chlorotoxin (CLTX) CARs and differentiated them into functional CAR-neutrophils. As compared to CAR-natural killer (NK) cells, systemically administered hPSC-derived CLTX CAR-neutrophils significantly reduced tumor burden in xenograft mouse models and extended their lifespan, suggesting superior abilities of neutrophils in crossing BBB and penetrating GBM xenograft in mice. We also loaded hypoxia-activated prodrug tirapazamine (TPZ) into CAR-neutrophils using silica nanoparticles with rough surfaces (R-SiO_2-TPZ) and demonstrated their enhanced antitumor activities in xenograft mouse models, serving as a novel dual chemoimmunotherapy against GBM. Our results established that CAR neutrophil-mediated drug delivery may provide an effective and universal strategy for specific targeting of solid tumors.

Nucs-Associated Mutation Avoidance in Mycobacteria and Its Association with Antibiotic Resistance

- Conference: International Conference on Stem Cell Engineering
- Year: 2022
- Proceeding: 8th International Conference on Stem Cell Engineering
- Group: General Submissions
- Session:

  Oral

Goblet cell differentiation in colorectal cancer-derived cell lines

Genetic and epigenetic changes provide survival advantages for stem or progenitor cells resulting in the deregulation of cellular differentiation. In colorectal cancer (CRC), this is reflected by the low frequencies of goblet cells. Understanding the mechanisms involved in goblet cell differentiation is, therefore, important for advances in the prevention and treatment of CRC. Our laboratory has a large collection of CRC cell lines, well characterised in terms of gene expression and mutations. We analysed the presence of goblet cells in our panel of cell lines using the genes Mucin 2 (MUC2) and Trefoil factor 3 (TFF3) as goblet cell markers. We found that most of the cell lines are unable to produce goblet cells and that the number of MUC2 and TFF3 positive cells among the goblet cell positive cell lines was quite variable. MUC2-negative, and TFF3-positive cell lines appear to reflect a novel interesting subset of CRC cell lines.

We also investigated the effect of transcription factors on goblet cell differentiation. Downregulation of Atonal homologue 1 (ATOH1) had a dramatic effect on goblet cell production in several CRC cell lines, while knocking down of SAM pointed domain ETS transcription factor (SPDEF), Caudal type homeobox 1 (CDX1), and 2 (CDX2) had a modest effect. Exogenous expression showed that, individually, none of these factors are sufficient on their own to trigger the differentiation of goblet cells. An improved understanding of cell differentiation will contribute to the treatment of colorectal carcinoma.

From iPSCs to Hematopoietic Lineage: An Endoribonuclease Mediated Cell-Type-Specific Genetic System to Guide Automated Differentiation

- Conference: International Conference on Stem Cell Engineering
- Year: 2022
- Proceeding: 8th International Conference on Stem Cell Engineering
- Group: General Submissions
- Session:

  Oral

Authors:
Wang, L. - Presenter, Institution of Process Engineering, Chinese Academy of Sciences, Beijing
Xu, W., MIT
Zhang, S., Massachusetts Institute of Thechnology
Weiss, R., Massachusetts Institute of Technology
hiPSCs provide unprecedented opportunities for cell therapies against intractable diseases. To produce therapeutic immune cells, recent advances emphasize the inevitable role of endothelial-to-hematopoietic transition (EHT), but still face a significant challenge in building a genetic program for cell-type-specific differentiation. To tackle this challenge, we built a novel platform for genetic programming, where endoribonucleases from the CRISPR family mediate sensing of endogenous cell-type-specific microRNA (miR) and actuate cell differentiation by transcriptional factor (TF) overexpression. We applied this system to differentiate hiPSC into hematopoietic lineages with an intermediate endothelial state. First, we extensively searched databases to identify endothelial-specific miRs and TFs that are critical for EHT. We designed and built high miR sensors and optimized the circuit elementsâ€™ stoichiometries by a poly-transfection method and achieved a fold change as high as 63. Then we verified the cell type specificity of this sensor using endothelial and non-endothelial cell lines. We monitored the temporal dynamics of endogenous miR during cell differentiation using a programmed differentiation from hiPSCs to endothelial cells via a 4-day protocol. With these solid validations, we replaced the reporter of the high miR sensor with TFs to guide differentiation to the hematopoietic lineage. Our results showed that the miR sensing and TFs actuating systems can derive significant percentages of CD34+CD45+ hematopoietic-like cells from hiPSCs. Together, this genetic system will provide a novel platform to program cell-type-specific differentiation, which will provide insights into the universal generation of functional and therapeutic immune cells from unlimited and reprogrammable hiPSC sources.

A Thermoresponsive Coating for Non-Enzymatic Release of Strongly Adherent Cells from Flasks and Other Closed-Vessel Formats

- Conference: International Conference on Stem Cell Engineering
- Year: 2022
- Proceeding: 8th International Conference on Stem Cell Engineering
- Group: General Submissions
- Session:

  Oral

Authors:
Garcia, E. - Presenter, ISurTec

Traditional cell culture techniques rely on enzymatic digestion for passaging, which introduces cost, cell stress pathway activation, potential contamination, and a reduction in cellular viability. With the damage perpetuated via repeated enzymatic-based harvests of cell lines, the hurdle for regulatory approval of stem cell and regenerative therapies increases. Thermoresponsive polymer coatings can reduce or eliminate the need for enzymatic digestion and improve cell line health. Current commercial options include coated petri dishes and multi-well plates, but no culture flask is yet widely available. Additionally, most of these products do not offer digestion-free passaging for strongly adherent cell lines such as mesenchymal stem cells (MSC). ISurTec has developed a thermoresponsive coating for T25 flasks that allows non-digestive cell release, optimized for strongly adherent cell lines. This coating will be expanded to other cultureware including larger and multi-layer flasks, and roller bottles, as well as dishes and multi-well plates. The thermoresponsive polymer is hydrophobic above 32Â°C and becomes hydrophilic at lower temperatures, spontaneously releasing the cells within 30 minutes when brought to room temperature. In comparison to standard culture flasks, our coating showed normal proliferation rates for multiple cell lines. Our coating releases 90% of human MSC (cord blood-derived) within 30 minutes as compared to 50% released on a competitorÃ¢â,¬â„¢s coating. Our coating also releases cells grown to 100% confluence instead of the 60-70% recommended limit of a competitor. ISurTherm can be applied to multiple cultureware substrates and is one of the only thermoresponsive culture flasks now available on the market.

Label-Free Three-Dimensional Morphological Analysis of Live Organoids Using Low-Coherent Holotomography

- Conference: International Conference on Stem Cell Engineering
- Year: 2022
- Proceeding: 8th International Conference on Stem Cell Engineering
- Group: General Submissions
- Session:

  Oral

Authors:
Lee, S. - Presenter, Tomocube Inc.
Lee, J., Tomocube Inc.
Lee, D. H., Division of Pulmonology and Critical Care Medicine, Department of Internal Medicine, College of Medicine, Chungnam National University
Lee, Y. J., Division of Pulmonology and Critical Care Medicine, Department of Internal Medicine, College of Medicine, Chungnam National University
Kim, H. J., Tomocube Inc.
Chung, C., Division of Pulmonology and Critical Care Medicine, Department of Internal Medicine, College of Medicine, Chungnam National University
Park, Y., Department of Physics, KAIST

Three-dimensional (3D) visualization and morphological dissection of organoids is challenging due to the thickness and its complex structure. Fluorescence-based imaging methods have been widely used; however, those are not suitable for observing morphological changes of growing organoids without damaging the samples. Holotomography (HT) has emerged as a useful tool for imaging live specimens without additional pre-treatment such as fixation, fluorescence labeling and excitation. HT can achieve long-term 3D observation of live specimens for weeks without phototoxicity. Furthermore, the measured individual cell data can be analyzed to elucidate not only the spatio-temporal 3D dynamics with high spatial resolution, but also the quantitative information of its subcellular compartments such as volume and dry mass.

In this study, we applied a recently developed low-coherent holotomography imaging system, HT-X1, which provides label-free 3D visualization of live cells and multicellular specimens. Using the HT-X1, we successfully observed *in situ* 3D morphological dynamics of patient-derived human lung organoids with Matrigel-embedded condition. The location and its thickness of the mucosal layer inside of which has a lower refractive index (RI) were also detected, since HT enables the volumetric reconstruction of RI which provides quantitative morphology analysis with unprecedented precision and resolution. The other subcellular features such as nuclei, nucleoli, mitochondria, lipid droplets, chromosomes, and multiciliate structures were observed with the high resolution of the HT-X1. Throughout the results, we suggest that low-coherent holotomography has an exclusive capability for phenotypic and quantitative studies of diverse types of organoids and other 3D multicellular live specimens including stem cells.

Extracellular Vesicles in Mesenchymal Stromal Cells Mediated Bone Regeneration

- Conference: International Conference on Stem Cell Engineering
- Year: 2022
- Proceeding: 8th International Conference on Stem Cell Engineering
- Group: General Submissions
- Session:

  Oral

Human mesenchymal stem cells (hMSCs) are widely used in regenerative medicine due to their differentiation abilities, secretion of trophic factors, and regulation of immune responses in damaged tissues. Due to the limited quantity of hMSCs isolated from bone marrow (BM), adipose tissue-derived hMSCs (hASCs) are considered a promising alternative. However, differences have been observed for hASCs in metabolic characteristics and response to *in vitro* culture stress compared to BM-hMSCs. In particular, the immune phenotype and immunomodulation of hASCs, remains unknown. This study thoroughly assessed the changes in metabolism, redox cycles, and immune phenotype of hASCs during *in vitro* expansion. In contrast to BM-hMSCs, hASCs did not respond to culture stress significantly during expansion as limited cellular senescence was observed. Notably, hASCs exhibited the increased secretion of pro-inflammatory cytokines and the decreased secretion of anti-inflammatory cytokines after extended culture. The NAD+/NADH redox cycle and other metabolic characteristics associated with aging were relatively stable, indicating that hASC functional decline may be regulated through an alternative mechanism rather than NAD+/Sirtuin aging pathways as observed in BM-hMSCs. Furthermore, transcriptome analysis by mRNA-sequencing revealed the upregulation of genes for pro-inflammatory cytokines and the downregulation of genes for anti-inflammatory cytokines for hASCs at high passage. Proteomics analysis indicated key pathways (e.g., tRNA charging, EIF2 signaling, and protein ubiquitination pathway) that may be associated with the immune phenotype shift of hASCs. Together, this study advances our understanding of the metabolism and senescence of hASCs and may offer vital insights for the biomanufacturing of hASCs for clinical use.

Modrna-Based CRISPR Systems for Robust Genome Editing in Human Pluripotent Stem Cells

- Conference: International Conference on Stem Cell Engineering
- Year: 2022
- Proceeding: 8th International Conference on Stem Cell Engineering
- Group: General Submissions
- Session:

  Oral

Authors:
Lian, X. - Presenter, University of Wisconsin-Madison

CRISPR systems have revolutionized biomedical research because they offer an unprecedented opportunity for genome editing. However, a bottleneck of applying CRISPR systems in human pluripotent stem cells (hPSCs) is how to deliver CRISPR effectors easily and efficiently. Here, we developed modified mRNA (modRNA)-based CRIPSR systems that utilized Cas9 and p53DD or base editor (ABE8e) modRNA for the purposes of knocking out genes in hPSCs via simple lipid-based transfection. ABE8e modRNA was employed to disrupt the splice donor site, resulting in defective splicing of the target transcript, and ultimately leading to gene knockout. Using our modRNA CRISPR system, we achieved 73.3% Â± 11.2% and 69.6 Â± 3.8% knockout efficiency with Cas9 plus p53DD modRNA, and ABE8e modRNA respectively, which was significantly higher than the plasmid-based systems. In summary, we demonstrate that our non-integrating modRNA-based CRISPR methods hold great promise as more efficient and accessible techniques for genome editing of hPSCs.

Goblet Cell Differentiation in Colorectal Cancer

- Conference: International Conference on Stem Cell Engineering
- Year: 2022
- Proceeding: 8th International Conference on Stem Cell Engineering
- Group: General Submissions
- Session:

  Oral

Authors:
Abdullayeva, G. - Presenter, University of Oxford
L. Lastun, V., University of Oxford
Bodmer, W., University of Oxford

Goblet cell differentiation in colorectal cancer-derived cell lines

Genetic and epigenetic changes provide survival advantages for stem or progenitor cells resulting in the deregulation of cellular differentiation. In colorectal cancer (CRC), this is reflected by the low frequencies of goblet cells. Understanding the mechanisms involved in goblet cell differentiation is, therefore, important for advances in the prevention and treatment of CRC. Our laboratory has a large collection of CRC cell lines, well characterised in terms of gene expression and mutations. We analysed the presence of goblet cells in our panel of cell lines using the genes Mucin 2 (MUC2) and Trefoil factor 3 (TFF3) as goblet cell markers. We found that most of the cell lines are unable to produce goblet cells and that the number of MUC2 and TFF3 positive cells among the goblet cell positive cell lines was quite variable. MUC2-negative, and TFF3-positive cell lines appear to reflect a novel interesting subset of CRC cell lines.

We also investigated the effect of transcription factors on goblet cell differentiation. Downregulation of Atonal homologue 1 (ATOH1) had a dramatic effect on goblet cell production in several CRC cell lines, while knocking down of SAM pointed domain ETS transcription factor (SPDEF), Caudal type homeobox 1 (CDX1), and 2 (CDX2) had a modest effect. Exogenous expression showed that, individually, none of these factors are sufficient on their own to trigger the differentiation of goblet cells. An improved understanding of cell differentiation will contribute to the treatment of colorectal carcinoma.

Modeling Amyloid-Beta Transport in hiPSC-Derived Models of the Blood-Brain Barrier

- Conference: International Conference on Stem Cell Engineering
- Year: 2022
- Proceeding: 8th International Conference on Stem Cell Engineering
- Group: General Submissions
- Session:

  Oral

Authors:
Eberts, P. - Presenter, University of Minnesota - Twin Cities
Azarin, S., University of Minnesota
Sarkar, C. A., University of Minnesota
One of the earliest detectable changes in sporadic Alzheimerâ€™s disease (AD) is dysfunction of the blood-brain barrier (BBB). A decrease in the ability of the BBB to transport amyloid-beta largely contributes to the accumulation of this peptide in the brain, a hallmark feature of AD. The causes of disrupted amyloid-beta transport across the BBB in AD are not fully understood. Here we used an *in vitro* model of the BBB composed of brain microvascular endothelial-like cells (iBMECs) differentiated from human induced pluripotent stem cells (hiPSCs) to model amyloid-beta transport. iBMEC expression of key receptors involved in amyloid-beta transcytosis, including the low-density lipoprotein receptor-related protein 1 (LRP1) and the receptor for advanced glycation endproducts (RAGE), was confirmed via immunocytochemistry and western blotting. Experimentation with iBMECs is commonly conducted 2-3 days after initial barrier formation. However, confocal microscopy revealed that proper polarization of the aforementioned receptors required iBMEC barriers to maintain a high transendothelial electrical resistance (TEER) (>1000 Î©Â·cm^2) for at least eight days after barrier formation. Polarized barriers exhibited greater rates of amyloid-beta transport and lower sodium fluorescein permeability compared to non-polarized barriers. To probe the changes observed in polarized versus non-polarized barriers, proteomic profiling was performed using 16-plex TMT-LC/LC-MS/MS. Principal component analysis and hierarchical clustering of these data revealed that polarized and non-polarized barriers have distinct proteomes. This collective work demonstrates the suitability of iBMECs for examining amyloid-beta transport and suggests that length of time in culture is a parameter that should be taken into account when working with iBMECs.

Effects of Physiological Relevant Bile Salts Combination on Duodenal/Ileum Monolayer: Mucin Secretion and Monolayer Integrity

- Conference: International Conference on Stem Cell Engineering
- Year: 2022
- Proceeding: 8th International Conference on Stem Cell Engineering
- Group: General Submissions
- Session:

  Poster Session

Authors:
Li, Y. - Presenter, Northeastern University
Carrier, R. L., Northeastern University
Bile acids/salts (BAs/BSs) are recognized as a family of stimulators that play essential roles in regulating various functions of physiology within the intestinal tract, like promoting the metabolism of dietary lipids, stimulating mucin secretion, and tuning the commensal bacterial community. However, many of the in vivo results were not able to be reproduced by in vitro models. And most in vitro studies were focused on single BA/BS treatments. Here, we set up a bile salts and phosphatidylcholine complex model (BS/PC) that mimics the human physiological intestinal luminal fluid at fasted state.

In this experiment, 8-9 days old duodenal and ileal monolayers were exposed to BS/PC for 24 hours. After the exposure, the permeabilities of monolayers and the mucin concentrations were measured by the Lucifer Yellow assay and the Alcian Blue (AB) assay, respectively. ZO-1 and MUC2 proteins on the monolayer were stained to characterize the abundance of tight junction and mucus.

According to the confocal imaged and AB assay data, BS/PC exposure altered the morphology of the duodenal monolayer and increased tight junction and mucus production of duodenal/ileal monolayers. The LY assay data showed BS/PC exposure did not affect monolayer permeability significantly, indicating thereâ€™s no damage of BS/PC to duodenal and ileal monolayers. The results indicated that the complex bile salts could affect the intestinal mucosa in vitro at a physiologically relevant concentration.

Gene Regulatory Network Analysis and Engineering Directs Development and Vascularization of Multilineage Human Liver Organoids

- Conference: International Conference on Stem Cell Engineering
- Year: 2022
- Proceeding: 8th International Conference on Stem Cell Engineering
- Group: General Submissions
- Session:

  Poster Session

Authors:
LeGraw, R. - Presenter, Arizona State University
Velazquez, J., Arizona State University
Moghadam, F., Arizona State University
Ebrahimkhani, M., University of Pittsburgh School of Medicine
Kiani, S., Arizona State University
Hislop, J., University of Pittsburgh Swanson School of Engineering
Cahan, P., Johns Hopkins University School of Medicine
Plaisier, C., Arizona State University
Chuva de Sousa Lopes, S., Leiden University Medical Center
Pluripotent stem cell (PSC)-derived organoids have emerged as novel multicellular models of human tissue development but display immature phenotypes, aberrant tissue fates, and a limited subset of cells. Here, we demonstrate that integrated analysis and engineering of gene regulatory networks (GRNs) in PSC-derived multilineage human liver organoids direct maturation and vascular morphogenesis in vitro. Overexpression of PROX1 and ATF5, combined with targeted CRISPR-based transcriptional activation of endogenous CYP3A4, reprograms tissue GRNs and improves native liver functions, such as FXR signaling, CYP3A4 enzymatic activity, and stromal cell reactivity. The engineered tissues possess superior liver identity when compared with other PSC-derived liver organoids and show the presence of hepatocyte, biliary, endothelial, and stellate-like cell populations in single-cell RNA-seq analysis. Finally, they show hepatic functions when studied in vivo. Collectively, our approach provides an experimental framework to direct organogenesis in vitro by systematically probing molecular pathways and transcriptional networks that promote tissue development.

Different Patterns of Chromatin Opening By Hematopoietic Pioneer Factors

- Conference: International Conference on Stem Cell Engineering
- Year: 2022
- Proceeding: 8th International Conference on Stem Cell Engineering
- Group: General Submissions
- Session:

  Poster Session

Authors:
Frederick, M. - Presenter, University of Pennsylvania
Zaret, K., University of Pennsylvania
Fernandez Garcia, M., University of Pennsylvania
Williamson, K., Harvard
Kadoch, C., Harvard
Luzete Monteiro, E., University of Pennsylvania
Ferretti, M., University of Pennsylvania
Pioneer transcription factors engage nucleosomal DNA in chromatin to initiate gene regulatory events that ultimately control cell fate. Pioneer factors can elicit local exposure of a nucleosome within chromatin and ultimately recruit co-regulators and remodelers to yield open chromatin sites seen in vivo. Yet how different pioneer transcription factors initially expose a targeted nucleosome in compacted chromatin structures is unclear. We used nucleosome arrays in vitro with a central nucleosome that can be targeted by the hematopoietic Ets factor PU.1 and the bZIP factors C/EBPÎ± and C/EBPÎ². Each class of factor can elicit targeted nucleosome exposure on linker histone-compacted arrays, but with different hypersensitivity patterns, as discerned from long-read nanopore sequencing. The DNA binding domains (DBDs) of PU.1 and C/EBPÎ± are sufficient for mononucleosome binding but are much less efficient than the full-length proteins, which can function cooperatively, in opening compacted chromatin. Thus, pioneer factors use their DBD to bind nucleosomes and other domains to elicit distinct patterns of target nucleosome exposure within compacted chromatin *in vitro*. Finally, nucleosome exposure by PU.1 enables further chromatin remodeling by a SWI/SNF complex. Together our data give a mechanistic view of how transcription factors cooperate to disrupt chromatin structures to initiate DNA accessibility to additional regulatory factors.

Synthetic Morphogenesis of Human Bilaminar Disc Via Genetically Encoded Transcriptional Programs

- Conference: International Conference on Stem Cell Engineering
- Year: 2022
- Proceeding: 8th International Conference on Stem Cell Engineering
- Group: General Submissions
- Session:

  Poster Session

Authors:
Hislop, J. - Presenter, University of Pittsburgh Swanson School of Engineering
Alavi, A., Carnegie Mellon University
Velazquez, J., Arizona State University
LeGraw, R., Arizona State University
Sozen, B., Yale University
Rytel, M., University of Pittsburgh
Mokhtari, T., University of Pittsburgh
Stolz, D. B.
Chuva de Sousa Lopes, S., Leiden University Medical Center
Kiani, S., Arizona State University
Bar-Joseph, Z., Carnegie Mellon University
Ebrahimkhani, M., University of Pittsburgh School of Medicine

Implantation of human embryo commences a critical developmental stage that comprises profound morphogenetic alteration of embryonic and extraembryonic tissues, axis formation and gastrulation events. Our mechanistic knowledge of this window of human life remains limited due to restricted access to natural healthy samples for both technical and ethical reasons. Here we describe a system that employs human induced pluripotent stem cells engineered with an inducible human transcription factor gene circuit to demonstrate genetically guided cell fates with single-cell transcriptional profiles similar to post-implantation embryonic and extraembryonic lineages. We show self-organization of these populations into three-dimensional multifate epiblast-like compartments surrounded by a yolk sac hypoblast layer. We observe self-organization and tissue boundary formation that recapitulates yolk sac-like tissue specification, the formation of bilaminar disc-like structure and the development of an amniotic-like cavity. This approach provides a simple platform for studying peri-implantation embryonic fate decisions and for the exploration of synthetic developmental engineering using a genetically encoded transcriptional program.

AUTHOR INDEX

AIChE
120 Wall Street, FL 23
New York, NY 10005-4020

ISBN 978-1-7138-6320-5